Summer 2001

Plastic *Passion*

Steve Nankervis

Schiffer Publishing Ltd

4880 Lower Valley Road, Atglen, PA 19310 USA

Designed by Bonnie M. Hensley
Cover Designed by Bruce M. Waters
Type set in Lydian BT bold/Zurich BT

ISBN: 0-7643-1213-8
Printed in China
1 2 3 4

Published by Schiffer Publishing Ltd.
4880 Lower Valley Road
Atglen, PA 19310
Phone: (610) 593-1777; Fax: (610) 593-2002
E-mail: Schifferbk@aol.com
Please visit our web site catalog at **www.schifferbooks.com**

This book may be purchased from the publisher. Include $3.95 for shipping. Please try your
bookstore first. We are always looking for people to write books on new and related subjects.
If you have an idea for a book please contact us at the above address. You may write for a free
catalog.

In Europe, Schiffer books are distributed by Bushwood Books
6 Marksbury Avenue Kew Gardens
Surrey TW9 4JF England
Phone: 44 (0) 20-8392-8585; Fax: 44 (0) 20-8392-9876
E-mail: Bushwd@aol.com
Free postage in the UK. Europe: air mail at cost.

This book is dedicated to
my wife, Maxine.

Acknowledgments

Photographs are by John Maclellan, except those as noted by Brandon McGuire.

The objects in this book belong to the author or John Maclellan and Jane Jardine, except those as noted to:

Astoria, 80 Rectory Grove, Leigh On Sea, Essex. SS9 2HJ. Tel. 01702-471800

Niven Casey

Violet Nankervis

Roger Stevens

Jim and Wendy Witherspoon

My thanks to the people mentioned above, and a very special thanks to John Maclellan for his excellent photography, hard work, and enthusiasm for this project.

Note:

Literkit® is a registered trade mark.

Values given in this book are current retail prices for objects in excellent condition at the time of writing. They cannot, however, be guaranteed.

Contents

Introduction

Plastics History

Plastics are materials that can be softened by heat and pressed into shape; amber and horn are two natural plastics.

The Naturals

Natural plastics have been used for centuries to make useful and beautiful objects. Of course, as the human race progresses we look for better and easier ways to do things. During the 19th century more natural plastics were utilised, one of which was shellac, made (unbelievably) from the secretion of a small insect found on acacia trees. This thermoplastic (plastic which can, when reheated, be re-used, as opposed to a thermoset plastic which, once set, keeps it's shape for life) was used to make gramophone records. Another natural plastic was Gutta Percha, made from tree bark and used to make jewellery, bottles, matchboxes and even furniture.

The Semi-synthetics

During the 1800s, advances in plastics science were made by combining chemicals and natural substances to create semi-synthetic plastics. Englishman Alexander Parkes invented Parkesine in 1855, which contained acids, oils, and wood flour; his wonderful mouldings were available from the early 1860s. Not long after this came celluloid. This remarkable thermoplastic was used to make such a vast array of items that it is hard to believe that they are made from the same material. Celluloid is a patented name for cellulose nitrate.

The Synthetics

Much scientific investigation in the plastics field was conducted at the beginning of the 20th century when the first synthetic plastic, Phenol Formaldehyde (Bakelite), was invented. This plastic was brittle and needed fillers to reinforce it, which give it the mottled look that is recognized today. The fillers restricted Bakelite to dark colors. After 1918, experiments with plastics really took off and many new plastics arrived on the market. Coupled with new moulding techniques, the new plastics enabled the industry to go from strength to strength. Thiourea Formaldehyde was prevalent during the 1920s, followed by Urea Formaldehyde which continued to improve plastics' capabilities. After 1940, another generation of plastics went into everyday use: vinyl, polystyrene, and nylon. And so progress continues.

Each decade of the twentieth century has brought advances in the plastics field. The next time you choose to play a CD (compact disc), take a closer look at the polycarbonate material it's made from and ask yourself where we would be without plastics. Now there's a frightening thought!

While looking at my plastics collection recently, my daughter asked, "Dad, when you're dead, can I have this." Puts it all into perspective, I think. Which brings to mind that a Bakelite coffin was the largest single Bakelite moulding. One can be seen at the Bakelite Museum in Somerset, England.

A Word in Praise of Plastics

Today, many people think of plastics as cheap and inferior, mainly due to mass produced goods that flooded the marketplace after the war of 1939-45. This is understandable as these toys, housewares and gifts were undeniably of poor quality. However, this was not always the case. The plastics industry of the late 19th and early 20th centuries (and I think to a certain extent today) aimed to make their products the best they could. Yes, they were mass produced, but often hand finished, beautifully designed and in colours that on mass can make you feel like you're a child in a sweet shop. I hope this book will invoke that feeling in you, as I have tried to include the most scrumptious objects available on the market.

Warning Before You Continue

I believe there are two types of people in the world: people who collect and people who don't. If you are a collector, read on with enthusiasm. If not, read on anyway for this book may make a collector of you yet.

A collector's time will be spent scouring antique fairs and shops for that next elusive plastics fix. No carboot (trunk) sale will be too early in the morning, no auction house too distant. You will spend time and more money than you ought, typically jus-

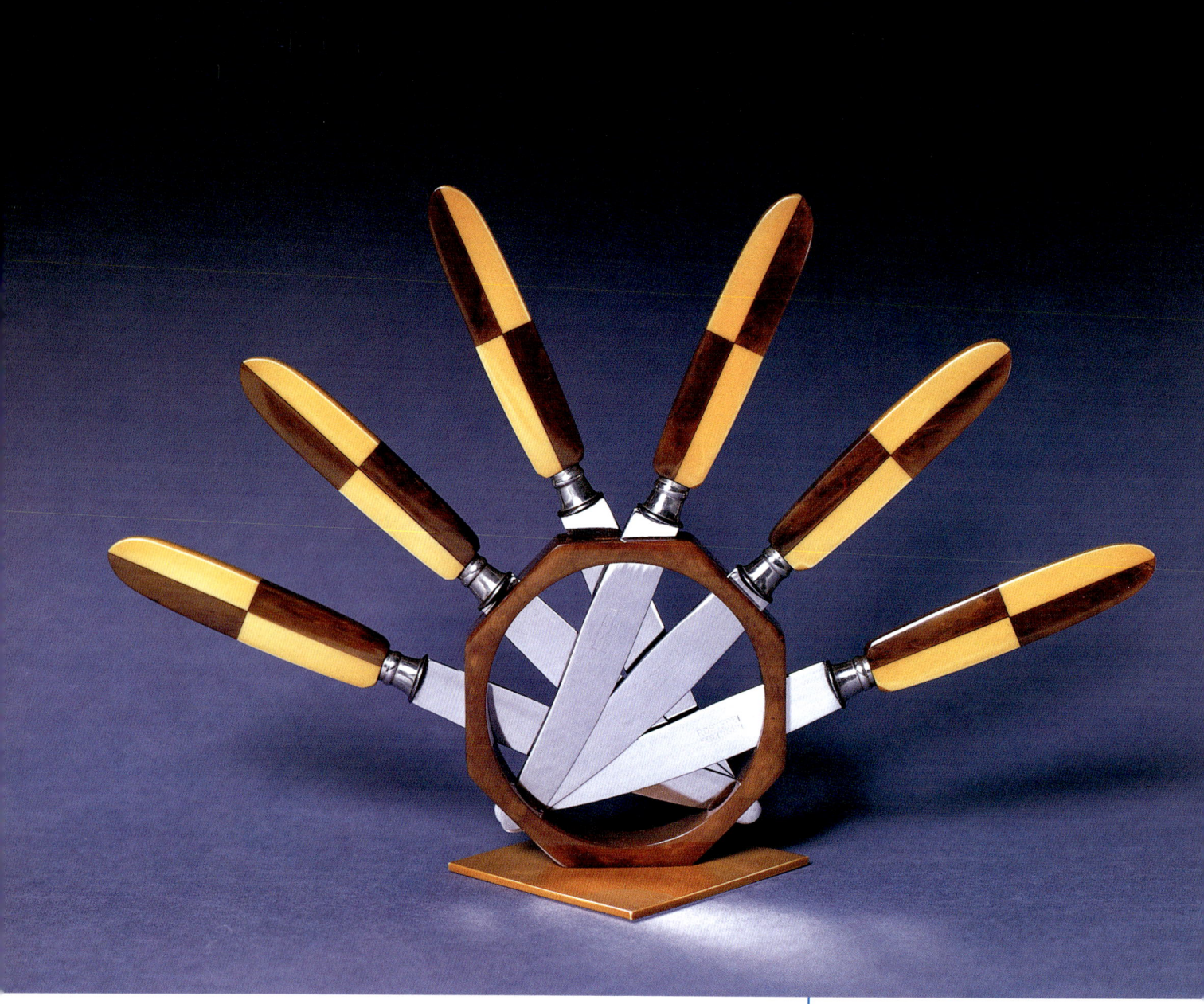

tifying your purchase with a casual, "I don't need to eat this week." You may then end up with a Bandalasta fruit bowl but no fruit to go in it. You may, of course, be made of sterner stuff than I. Good luck.

Chapter 1.

"Pop It On The Sideboard"

Plastics in the Home

Here we look at objects for the home. Part of the joy of collecting is not only to have a group of objects on display, but to also have things dotted about the house. Of course, you have to work hard to find these things due to their rarity. However, the reward for this effort is owning things that help to give your home individuality and style. When looking for a photoframe for that special holiday snap, most people will go to a department store - much better then to make a little extra effort and get yourself a carvacraft photoframe which not only looks great, but will quickly increase in value. So clear the mantlepiece for homeloving plastics…

Animal Napkin Rings made of cast phenolic during the 30s and 40s. Could be purchased individually or in sets. Also available in an extruded rod form to cut up yourself. $30-40.

OPPOSITE PAGE BOTTOM
Pachyderms on parade. Napkin rings. $30-40 each, $80-90 with wheels.

Red and green duck with contrasting colored eyes. These were drilled and a tube of resin inserted. $30-40 each.

13

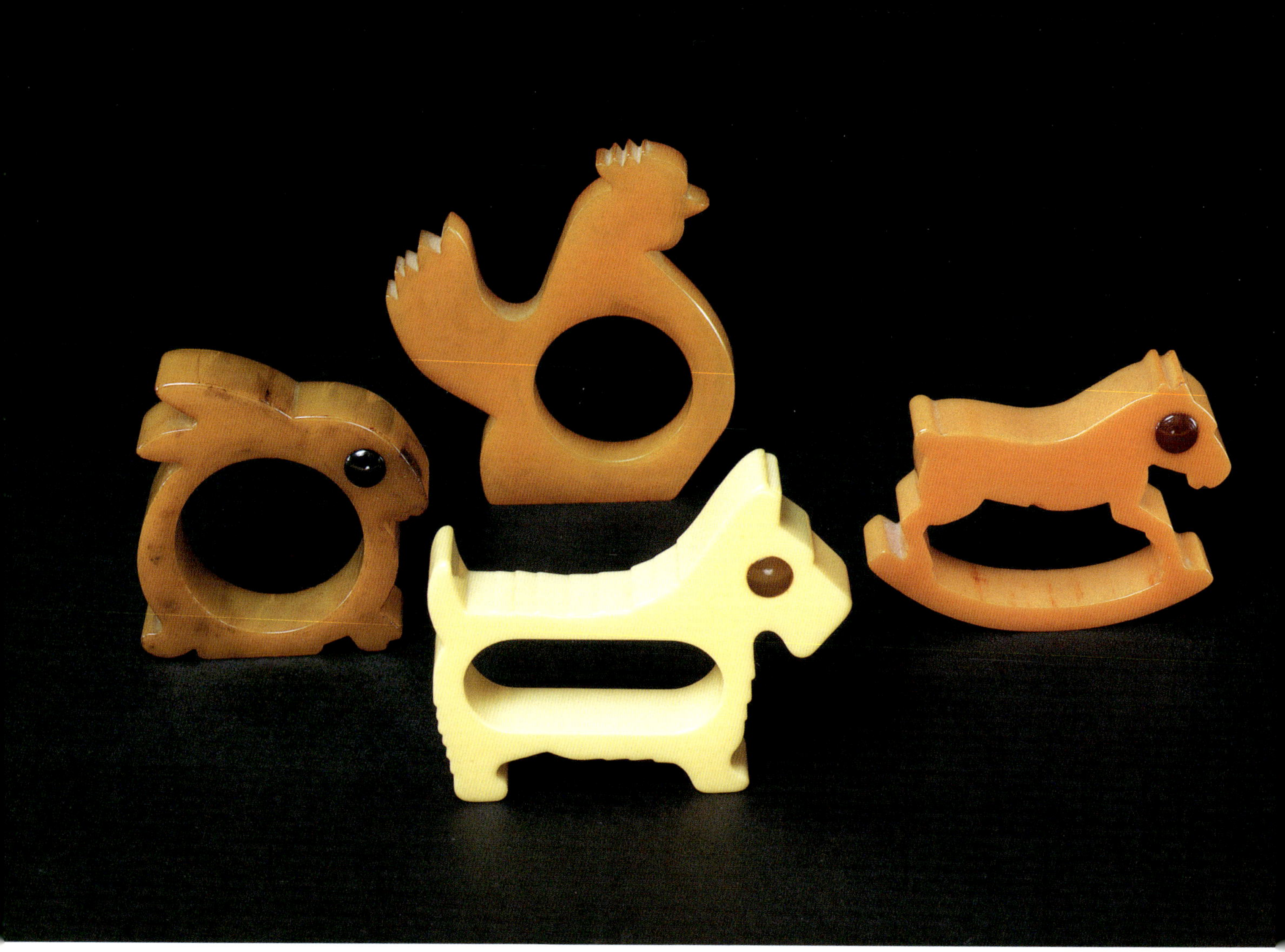

Camel, "Of course I'm the rarest of the Napkin Rings." Horse, "I think you'll find that's me actually." *Camel Courtesy of Astoria*. $110-130 each.

Phenolic birds with wooden rings. $110-130 set of 6.

Bakelite- the first synthetic plastic, Phenol Formaldehyde, which was discovered by Belgian chemist Dr. Leo Baekeland while he was working in New York in 1907. Phenol Formaldehyde was patented in 1909 and known as the "heat and pressure patent." He then formed The General Bakelite Company in 1910. The new product, called Bakelite, was a big success and was used extensively in industry and the home. Bakelite has now become a generic term and is used to describe many types of plastics. Antique dealers use the word "Bakelite" to mean dark mottled and marbled plastics and the word "phenolic" to describe the more translucent, brighter colors. This is not always technically correct, but it works in the marketplace.

Phenolic Napkin Rings. Sets were once common but are now becoming scarce. $45-55 set of 6.

Pair of Napkin Rings, Celluloid on Plaster. $55-65 each.

Phenolic- **A shortened version of "Phenol Formaldehyde," the first synthetic plastic. Previously, "plastics" had been natural (horn, shellac) or semi-synthetic (celluloid, casein).**

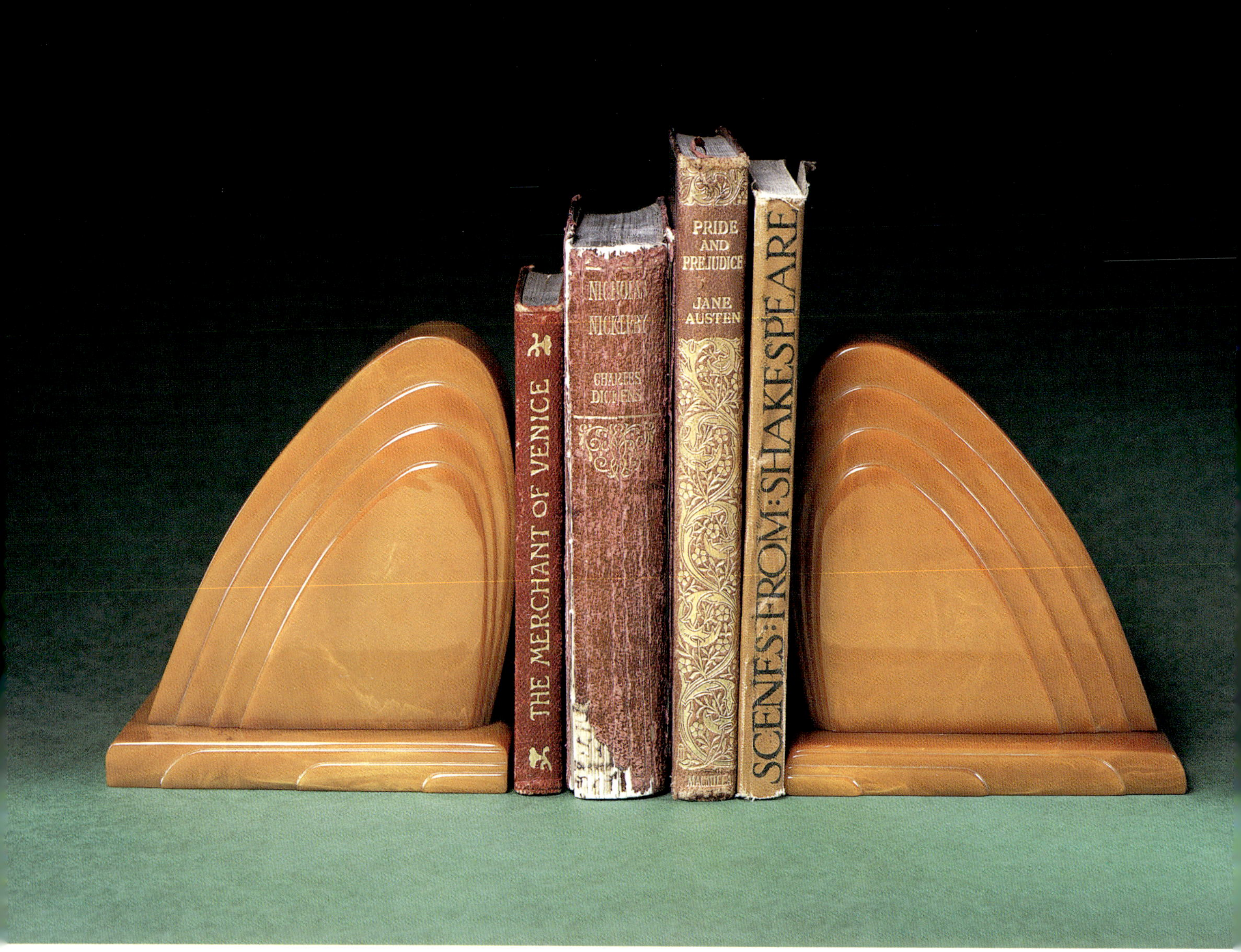
THE MERCHANT OF VENICE
CHARLES DICKENS
PRIDE AND PREJUDICE
JANE AUSTEN
SCENES:FROM:SHAKESPEARE

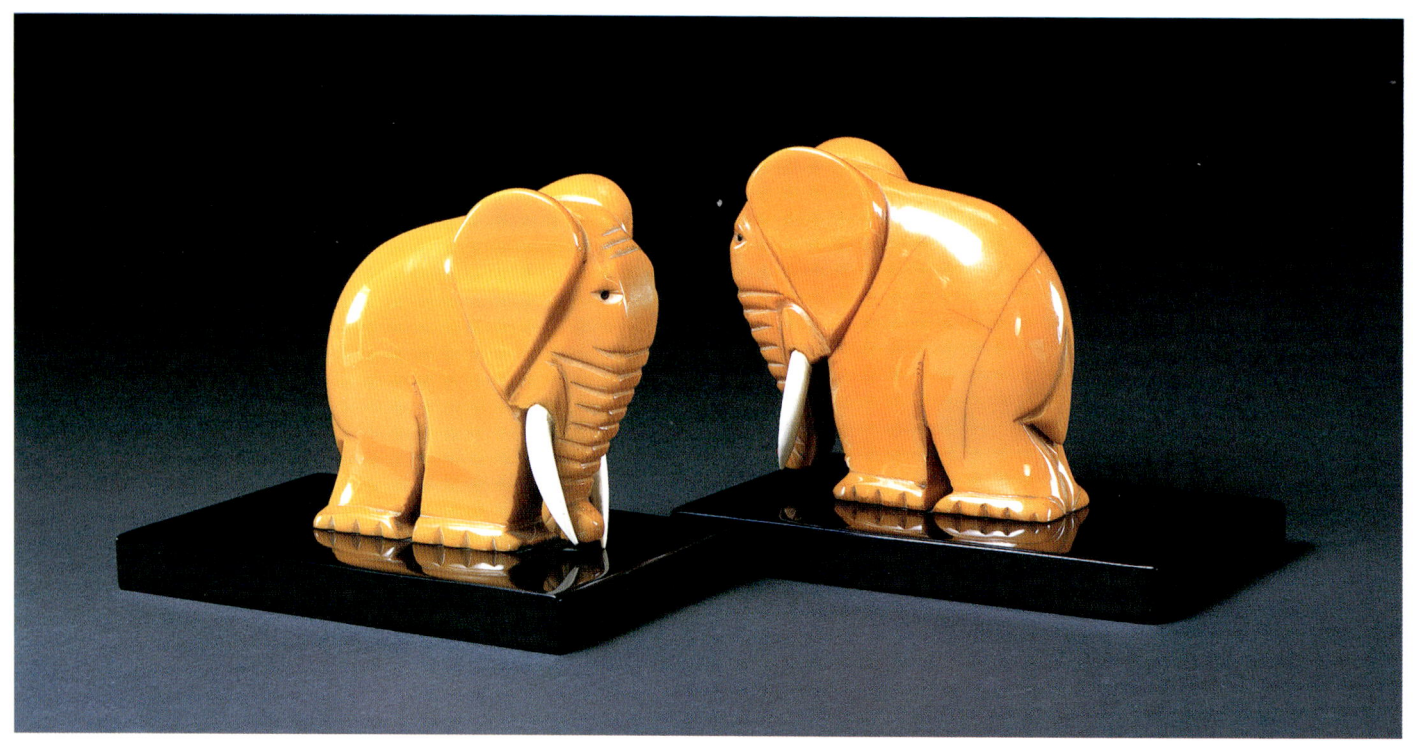

Stylized bird. Possibly one of a pair. Phenolic, probably 1930s. *Courtesy of Astoria*. $110-125 each.

OPPOSITE PAGE

TOP
Carvacraft Bookends. Part of a desk set designed by Charles Boyton for Dickenson, Hemel Hempstead, in the late 1940s and '50s. Available in green, amber and yellow phenolic. $450-$600.

BOTTOM
Elephant Bookends. Phenolic, 1930s. *Courtesy of Astoria.* $550-$700.

Calendar in red phenolic. 3" high.
$60-70.

Calandox Calendar. Bakelite with celluloid face. *Photo by Brandon McGuire.* $45-50.

Carvacraft photo
frame. 3.5" high.
$120-140.

Phenolic photo
frame. 7" high.
$100-120.

Acrylic photo frame. Base is
chrome on brass. c.1970. $45-50.

Acrylic- **Plastic glass, better
known by ICI's trade name
Perspex and Lucite in America,
available from the 1930s onward.**

Photo frame. Two separate arms slot
into the base. Phenolic. 1930s. $35-40.

Pen Holder. Phenolic.
1930s. $70-90.

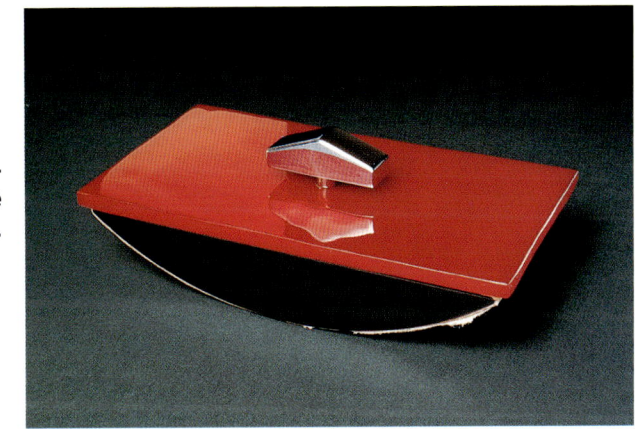

Red and black phe-
nolic blotter, chrome
handle. $40-55.

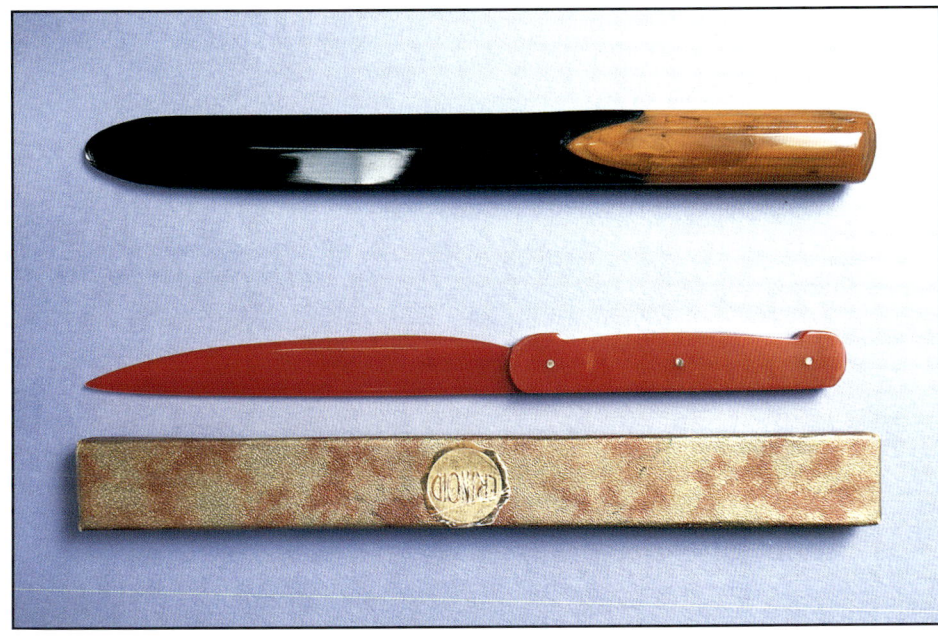

Top. Carvacraft paper knife. Was
also available with a clear acrylic
blade, 8" long. $45-50.
Red paper knife with original box
marked Erinoid. $55-60.

Erinoid- **Erinoid Limited was a company which produced Casein, a semi-synthetic plastic which contained milk protein. The name has become a generic term. Erinoid was successfully used to make many items, including knitting needles.**

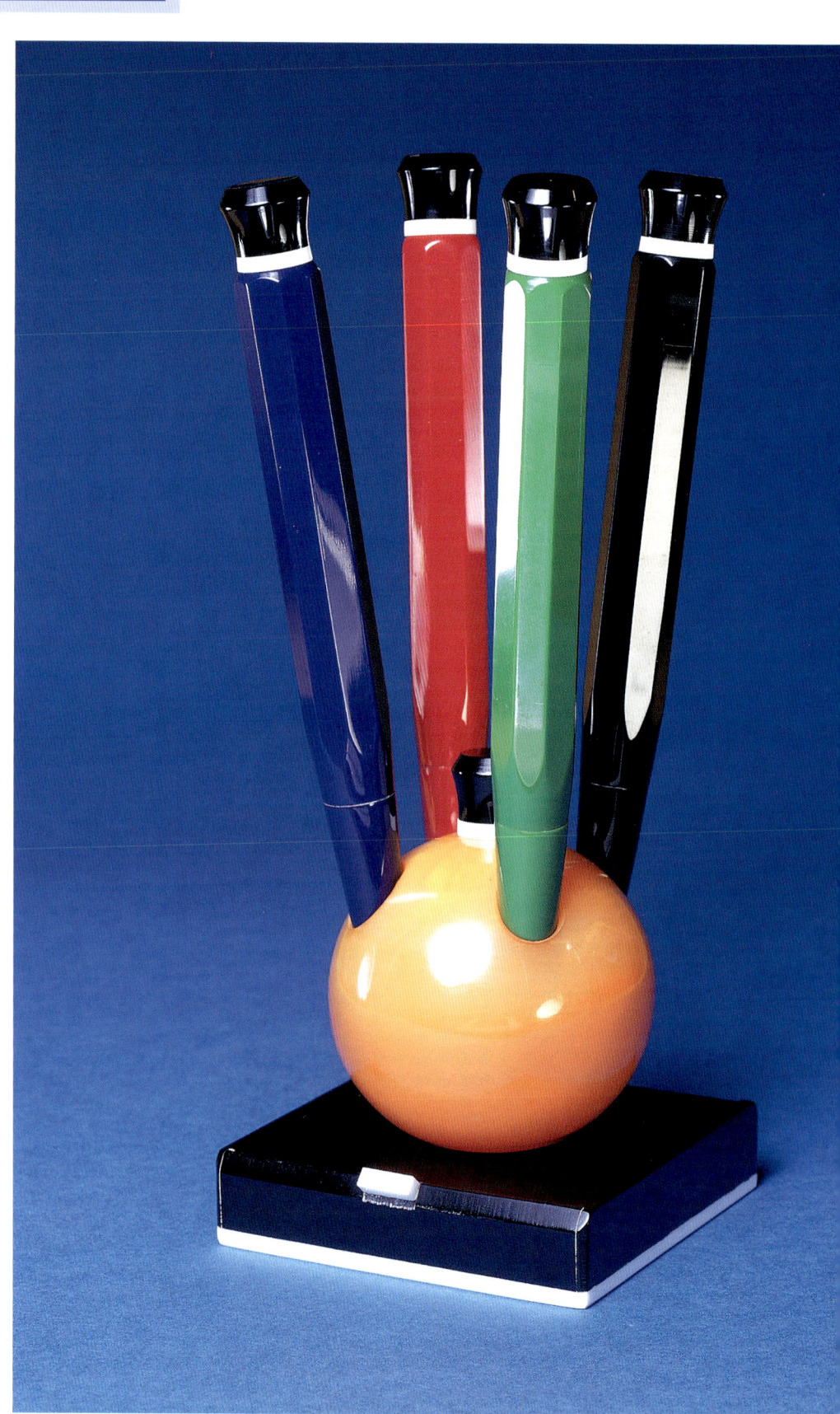

Phenolic pen set. Base slides back to reveal spare leads. 7" high. $250-300.

332 Series telephone, the first to have an integral bell box. Acrylic, also made in urea formaldehyde from the 1930s onwards. Has been reproduced, so buy with care. $300-325.

Smart Deco styled barometer, 5" high. Made in England in the 1930s. Chrome and urea formaldehyde. The photograph was taken in Great Britain during one of the rare occasions when the needle was pointing to very dry. $40-50.

Rototherm thermometer in green urea formaldehyde. Available in other colors. Includes a central comfort zone, supposedly to let you know when you're comfortable. The ice says cool, the lights say hot - the lights won. *Photo by Brandon McGuire.* $35-40.

Electric Clock made in England by Hawkins. In dark red and black mottled Bakelite sheet and amber phenolic. 8.5" high. $125-145.

Green marbled phenolic clock made by Goblin. 5.5" high. $85-95.

Smiths electric clock. The best feature of which must be the brown and cream urea formaldehyde in which it is finished. English, 1950s. $30-40.

Multicolored clock
molded by Streetly in
urea. 1931. $70-80.

Phenolic clock, 1940s.
Courtesy of Niven Casey.
$85-95.

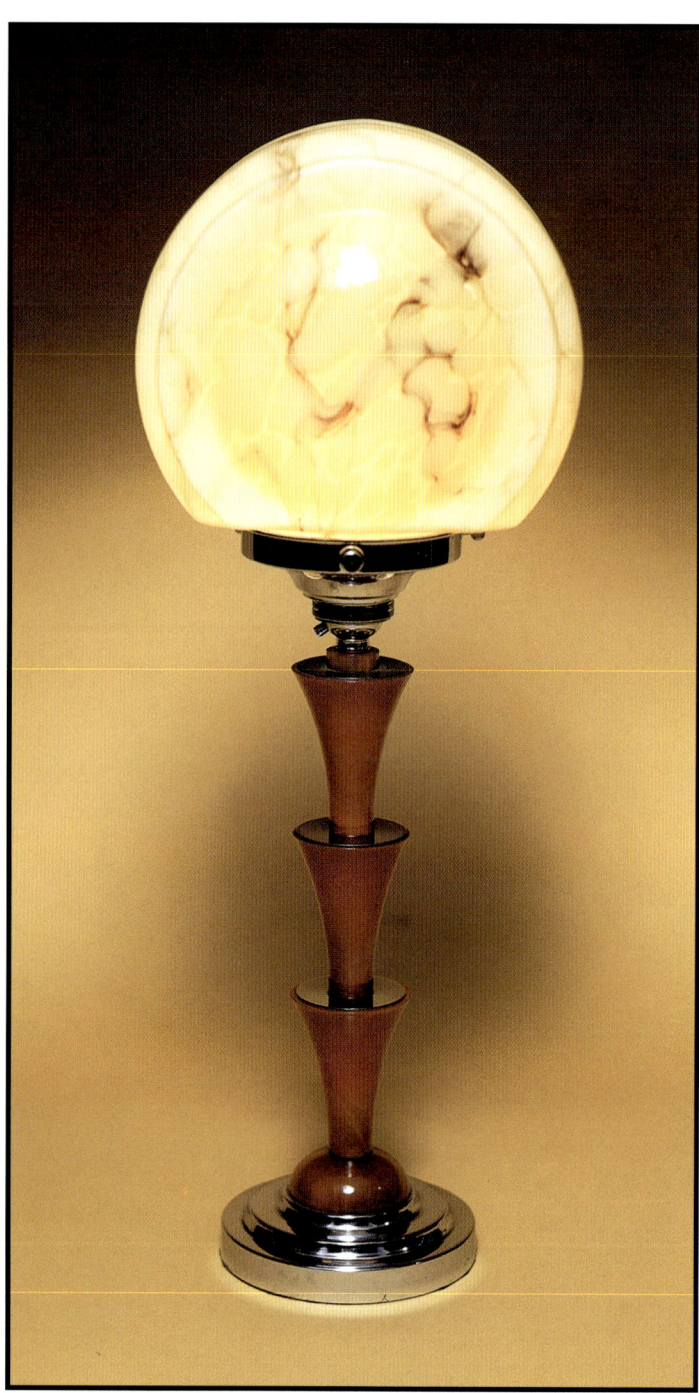

Phenolic lamp, each trumpet topped with a chrome disc. 17" high. 1930s. $200-240.

Celluloid- **Invented by John Wesley Hyatt in the late 1800s. A semi-synthetic plastic containing powdered camphor. A big success, it's uses included piano keys, dolls, dentures, shirt collars and of course, cine film. It's major drawback is that it is flammable.**

OPPOSITE PAGE
Celluloid lamp, probably 1930s. The celluloid sheet is stitched to a wire frame. It measures over 2 feet high. Likes a shady spot away from sunlight. $600-650.

Deco table lamp. Acrylic sheet base with ebonite step. The ebonite has turned brown due to age. $200-250.

Phenolic lamp created by using sections from a standard or ceiling light. These are easier to sell. 16" high. *Courtesy of Roger Stevens.* $140-160.

A selection of light switches, phenolic and pearlized celluloid, 2nd from right. $15-20 each.

Aladdin lamp by Pifco. Lights when picked up. 3" high. Stamped Empire Made. $20-25.

Bristly chests reveal the use of these gossiping ducks, they are in fact clothes brushes. PVC with a nylon brush. 11.5" high. A must for the well turned out collector. $20 each and $30 each for blue, red, grey, and green examples.

Nylon- **A synthetic first shown at the New York World's Fair in 1939 as bristles for tooth and hair brushes and, of course, nylon stockings.**

Phenolic yacht. Possibly Polish, 1930s. 4" high. $55-70.

What a great place to hide your treasure. Metal spine and brackets with phenolic cover. $95-105.

Chrome and phenolic candlesticks. 4" high. 1930s. $75-80.

Candlesticks in phenolic and chrome. $85-90.

Candlesticks. Phenolic and chrome. $175-200.

Bournvita mug from Cadbury's Chocolates. This wonderful mug was made in melamine and polythene. This one dates from 1951 and is urea formaldehyde. $120-135.

Bakelite fruit bowl. Linsden Ware, made in England. 1930s. *Photo by Brandon McGuire.* $45-55.

Beatl fruit bowl by Streetly. *Photo by Brandon McGuire.* $200-250.

Beatl- Products made from beetle powder were displayed in Harrods in 1926 and sold by Selfridges soon after. Originally called Beetle, but this was changed to Beatl after complaints from customers who did not like the insect connotations. Note no such objections were heard from Woolworths, who also stocked this tableware.

Acelloid ware bowls. Made in England. $30-35 each.

Melamine- **Often known as Formica or Wareite, it was used as tableware by US Airlines and kitchen worktops in England during the 1950s.**

Fruit knife. Ivory, bone and celluloid. c.1900. $50-55.

OPPOSITE PAGE
A lovely set of six knives in phenolic. $125-145.

Phenolic knife set. $120-135.

Cake forks and slice. $80-95.

40   Knife sets. Made in Sheffield, England. $75-80 per set.S

Egg cup and salt set in Bakelite, urea and acrylic. 16 separate pieces. What good quality. Expect to pay around $22-32.

Cruet set in Bakelite. $15-20.

This very camp Heatmaster teapot has a thermal liner to ensure your second cup is as hot as the first. English, 1940s and '50s. *Courtesy of Jim and Wendy Witherspoon.* $75-85.

Bakelite juicer. *Courtesy of Jim and Wendy Witherspoon.* $30-40.

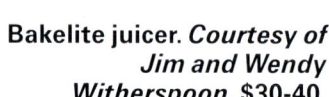

Green and white
marbled flask by
Thermos. Late 1930s.
Urea formaldehyde.
$30-35.

Urea Formaldehyde- **Invented in the 1920s, this**
helped expand color variations including white,
not before available. Brookes and Adams pro-
duced it's Bandalasta and Linga Longa ranges
using Urea and Thiourea Formaldehyde.

The answer to all our transport problems? No, it's the Smoothie travel iron from the late 1940s. $22-32.

Ribbed phenolic door handles. 7" long. $35-40 pair.

Impress the post-man with a Bakelite letterbox. Possibly 1930s. $30-40.

Chapter 2.

"Light Me A Gasper And Pass The Bolly"

Smoking and Drinking

During the 1920s, '30s and '40s, everyone, it seemed, wished to live a hedonistic lifestyle. Advertisements would show fashionable ladies and gents smoking and drinking themselves into a giddy never ending night on the town. The new plastics with their bright colours, cheap mass production and availability could be made into ashtrays, lighters, cigarette holders, cocktail favours etc. Companies were quick to use them to help promote their brands, often as giveaways (50 State Express cigarettes in a decorative urea box has much more appeal than the usual cardboard carton) so in this chapter we pick up our Luckies, grab a Gilbeys and tonic and remember a time when alcohol was purely medicinal and cigarettes could cure a sore throat.

Photo by Brandon McGuire.S

Cigarette box. Mid-1930s. Urea formaldehyde. $75-80.

Grecian dancers adorn this urea formaldehyde cigarette box. Molded by Streetly and marked State Express. Mid-1930s. Note the strategically placed tambourine. *Courtesy of Roger Stevens.* $40-50.

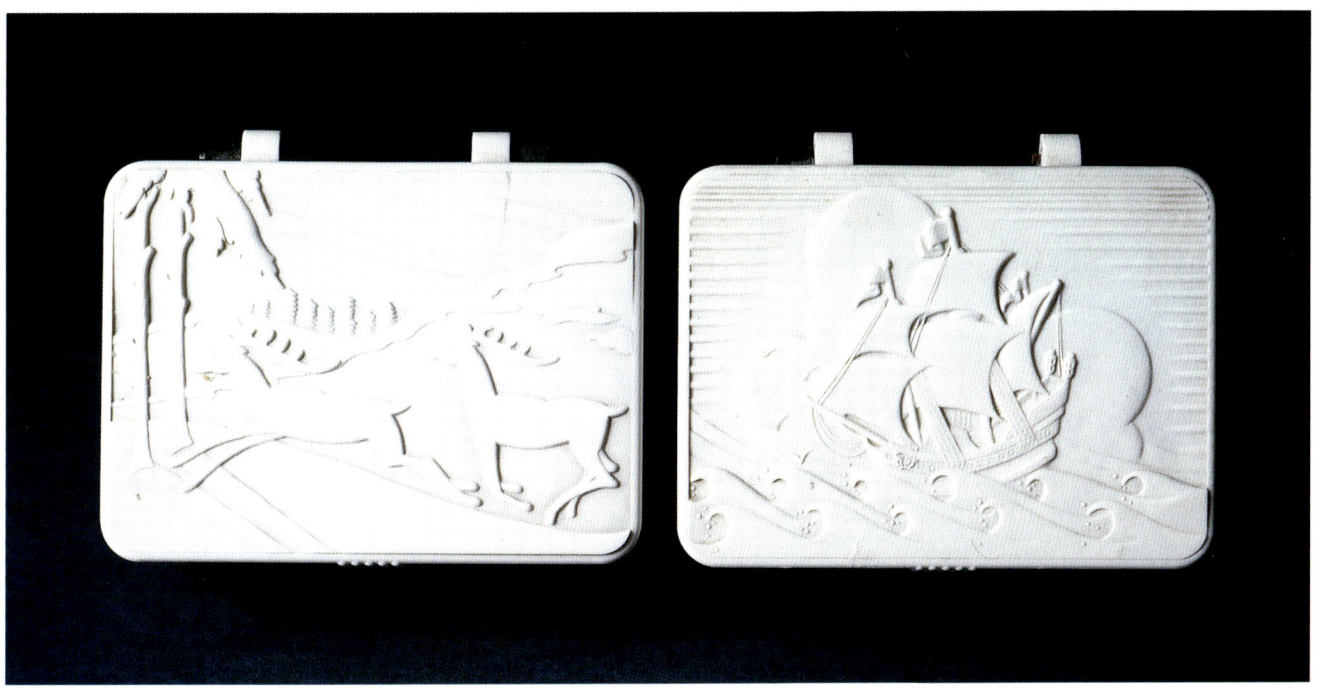

Two more similar boxes, less desirable. 1938. $15-20 each.

Yellow phenolic box marked The British Buttner Pipe Co. Ltd. 3.25" high. $85-100.

T'was indeed a terrifying sight! This particular beastie has an ashtray in his tail and a striker in his mouth. Made by the YZ Novelty Company for Dunhill in the 1920s and '30s. Phenolic. $200-250.

Sid and Johnny have phenolic beaks and sharkskin strikers. $110-120 each.

Bird made by the YZ Novelty Company. How many different ones were produced is unknown, as records were lost in a fire during World War Two. Phenolic head, beak, and feet; missing ashtray. $100-115.

Caterpiller is all phenolic and very rare. 1930s. $240-300.

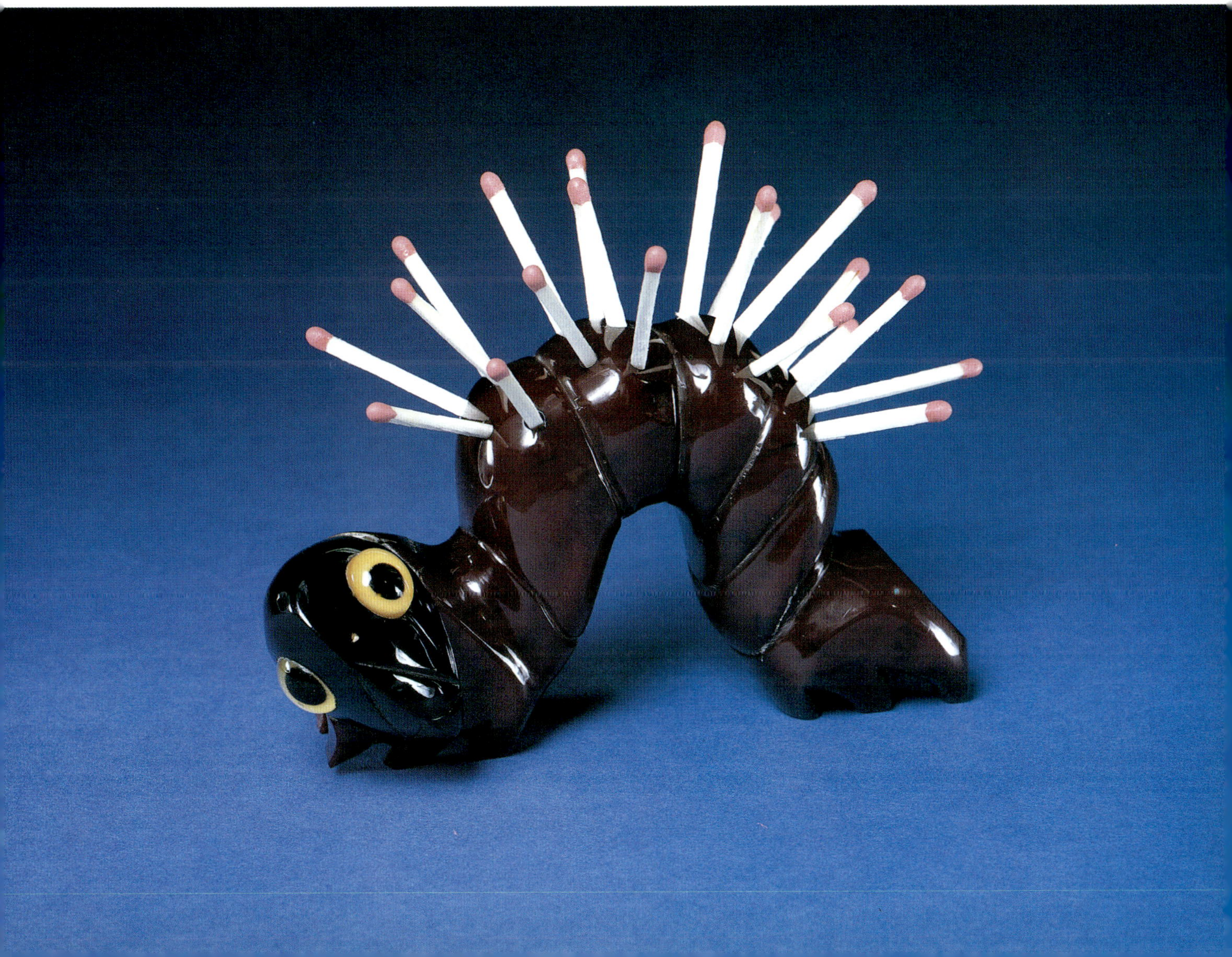

Frank (Ol' Blue Eyes) is barely 2" tall, but it still manages to command a price of around $105.

Bird made by YZ Novelty Company. $95-115.

Baby bird ashtray.
Phenolic. $60-80.

Mixed mediums bird made by YZ
Novelty Company in wood, marble,
and phenolic. $75-80.

The Roanoid ashtray made for Dunlop, who gave them away to their customers. Available in several colors. They are wonderful objects. 1930s. $300-325 each.

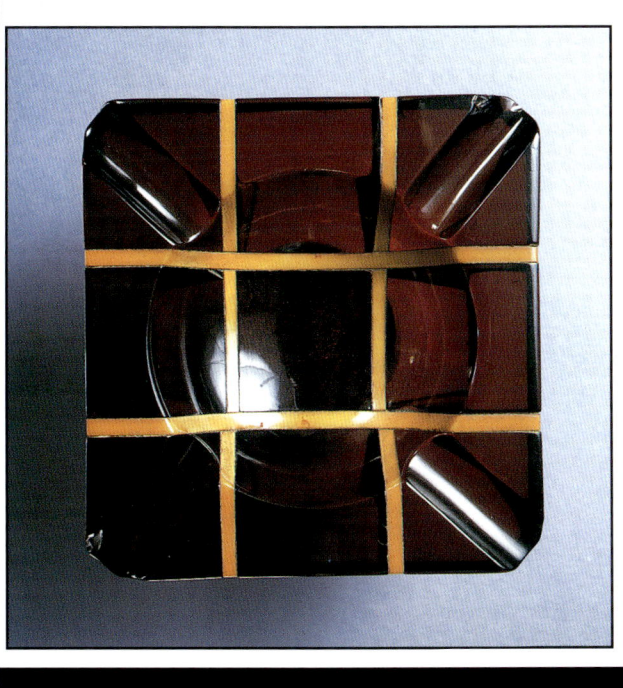

Phenolic ashtray.
$60-80.

Souvenir ashtray of the
Atomium, Brussels. Melamine.
1950s. $30-35.

Red and black mottled Bakelite heart conceals a petrol lighter. 2" high. $110-125.

This lady's handbag accessory contains a cigarette holder and solid perfume. Phenolic and celluloid with handpainted cherry motif. $70-75.

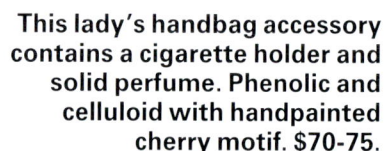

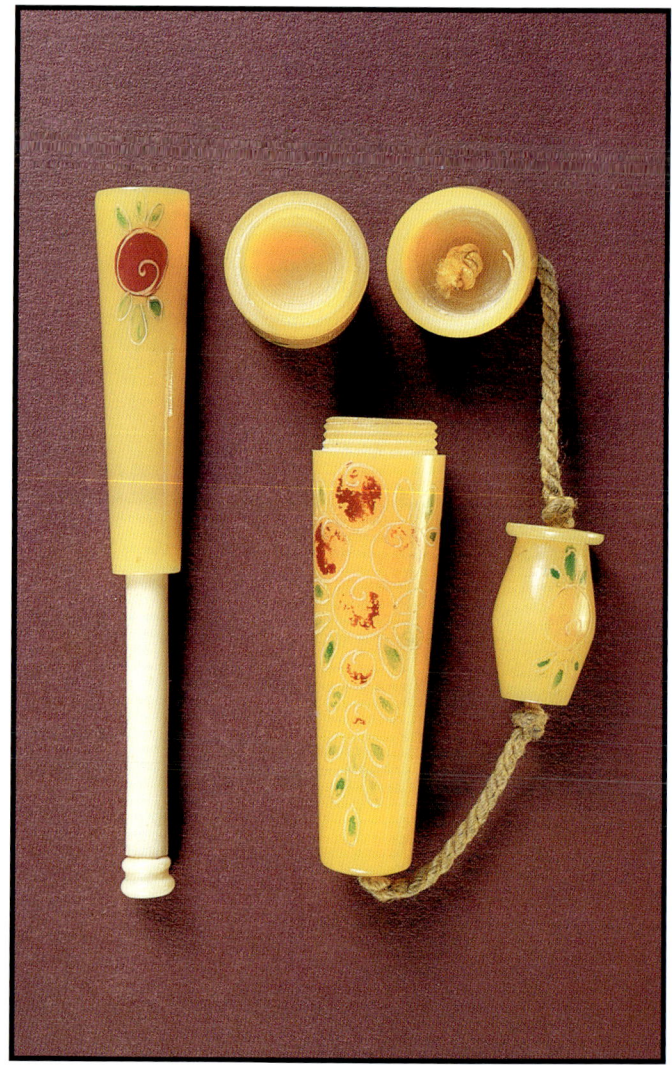

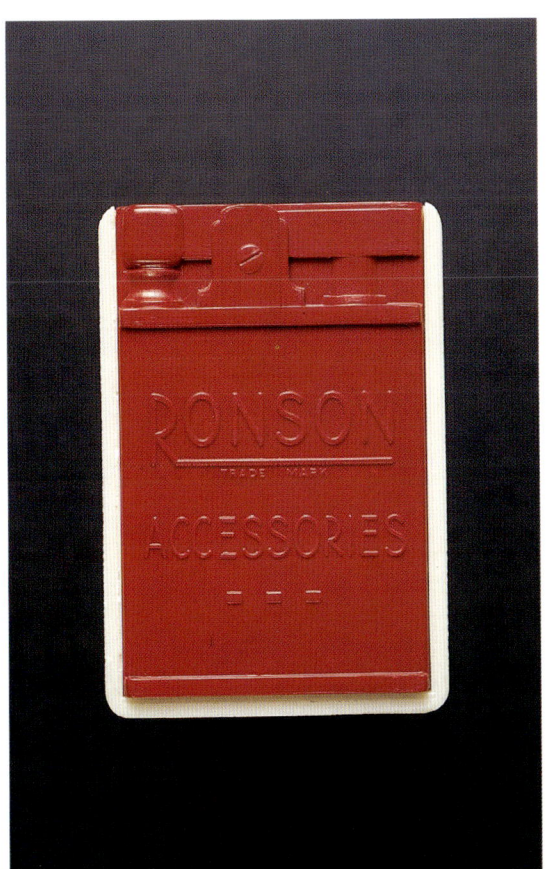

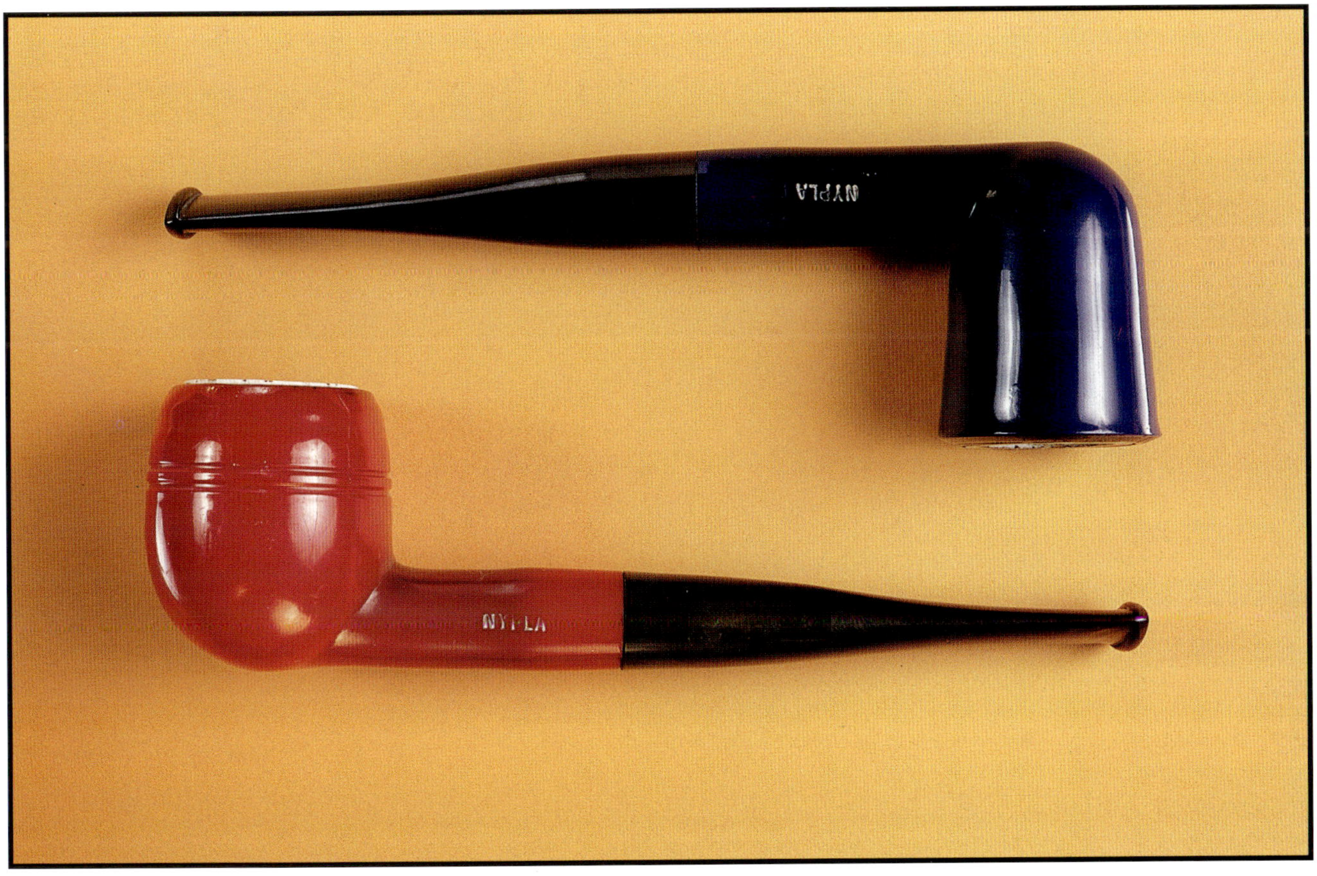

Live in peace with your pipe. Phenolic. About $40 each.

Celluloid cigarette case. 1920s. Difficult to find in perfect condition. 3" high. $90-110.

Indian cigarette holder with combined ashtray. The cigarette is placed in the Indian's mouth, ensuring you never suffer the embarrassment of dropping ash on your friends' Betty Joel rug. 1930s. $50-60.

Bottle opener. A popular collectible in metal and cellulose acetate. English. 1930s. $110-120.

Cellulose Acetate- **A 1920s plastic which became a replacement for celluloid as it was non-flammable. Still in use today.**

This decorative item is very well made in acrylic; the ashtray actually contains some plastic ash! 3.75" high. Date and origin unknown. $80-125.

Cocktail marker. Celluloid. Possibly 1920s. $140-150.

Cocktail marker lighthouse, shipswheel and sailors, whose heads sit on the edge of a glass. Phenolic. 1930s. $140-170.

Chapter 3.

"Drop Dead Gorgeous"

Jewelry, Perfume, and Accessories

This area of collecting plastics seems almost recession proof. When times are good people treat themselves to some jewelry, when times are bad people cheer themselves up by treating themselves to some jewelry. Sources are not infinite, so prices rise. This trend will continue to make jewelry a consistently good investment unless the market is spoiled by a flood of reproductions. So ladies, pop on some lippy and gentlemen, a dab of cologne; get ready for the absolutely fabulous Chapter 3.

Evening In Paris perfume presentation
mottled phenolic. The shoes are cellulc
Molded by Prestware in the late 1930s
author's favorites. $180-200.

Evening In Paris perfume Owl
presentation box. Phenolic with
glass eyes. 3.75" high. $160-180.

Evening In Paris perfume Shell presentation box. *Courtesy of Jim and Wendy Witherspoon.* $140-160.

Top Hat presentation box in urea formaldehyde. Molded by Streetly for the Saville Perfumery Company. It cost 1s 9d (approx. 9p) in 1937. Now expect to pay around $80.

Phenolic perfume bottle. 4" high. $70-90.

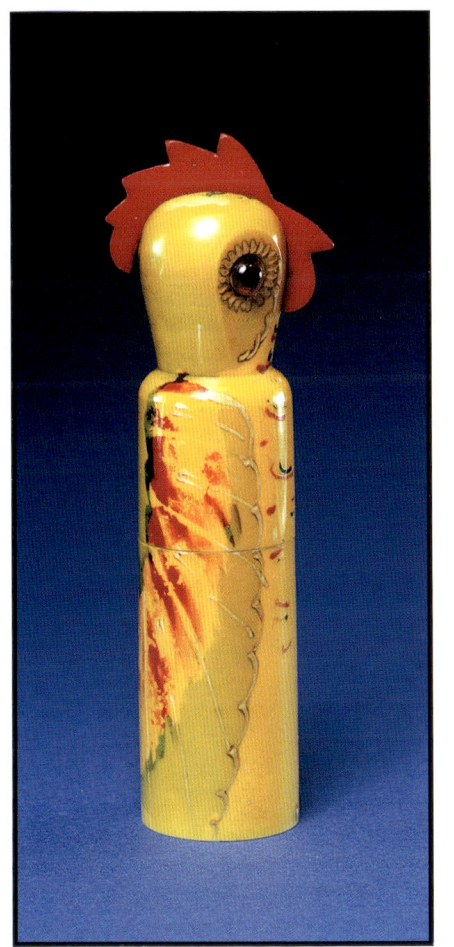

Dice which contains solid perfume. Still smells OK, albeit with a hint of Bakelite. 0.75" high. $55-70.

Perfume bottle with glass liner. $80.

Perfume bottle with
glass vial. $110-120.

Boxed set contains perfume,
lipstick, powder and solid
perfume. Phenolic with hand
painted decoration. 1930s.
$350-400.

Celluloid vanity set, late 1920s. German. $140-200.

Powder box with blue
swirl top and feet.
$70-80.

Box by Fornells, Paris. The gold paint has
been applied over the urea to accentuate
the decorative relief. The inside is also gold
and states Fornells Editions Paris. This
company's products are highly prized.
Possible late 1920s or early '30s. 3.6" high.
$240-300.

66

Phenolic box. Part of a
dressing table set. $80-90.

Box from a dressing table
set. 1930s. $50-60.

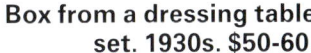

Powder box. Celluloid. $70-80.

Powder box. Urea
formaldehyde. 1940s.
$20-25.

Art Deco dressing table set in a mixture of
plastics. Includes tools for nail care. Note fading
to right side caused by partial exposure to
daylight. Probably French, 1930s. *Courtesy of
Astoria.* $400-475.

Powder puff. Multi-colored celluloid, 1930s. $225-280 with original box.

Powder puff, celluloid 1920s. $225-280 with original box.

French powder puffs which open and close with a
wonderful telescopic action. The puff is swan's
down. Phenolic. $70-90 each.

Powder compact, phenolic
1930s. $70-80.

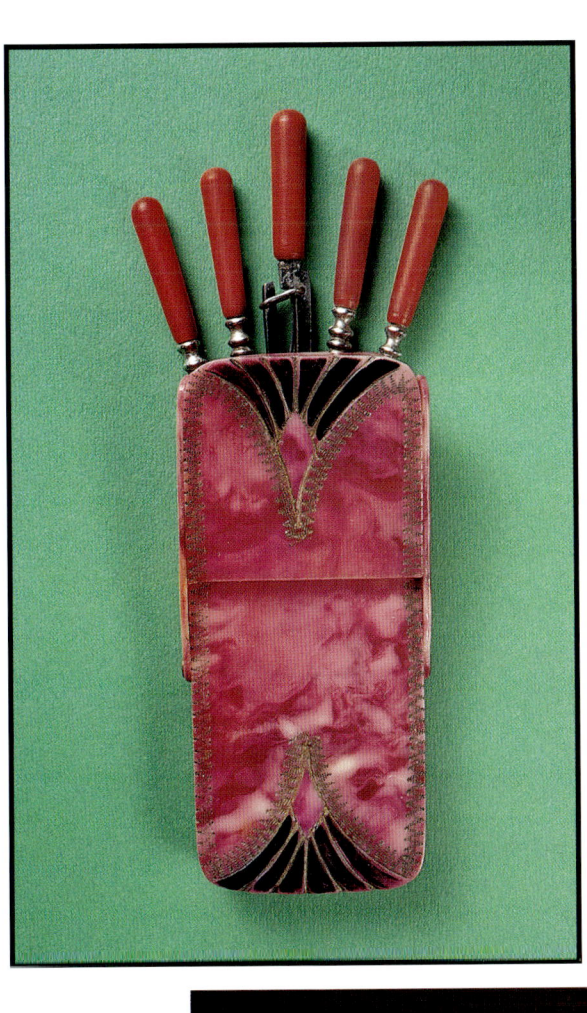

Manicure set in celluloid
case. $50-60.

Ribbed phenolic and
acrylic box. $35-40.

How do you like your eggs, fried or boiled - or made of celluloid? Jewelry cases. About $80 each.

LEFT
Elasticated bangle, metal spacers. $20-25.

RIGHT
Spark plug cuff links in urea formaldehyde. $50-55.

LEFT
Phenolic brooch. It's the berries! $85-90.

RIGHT
Celluloid flower basket brooch. $35-40.

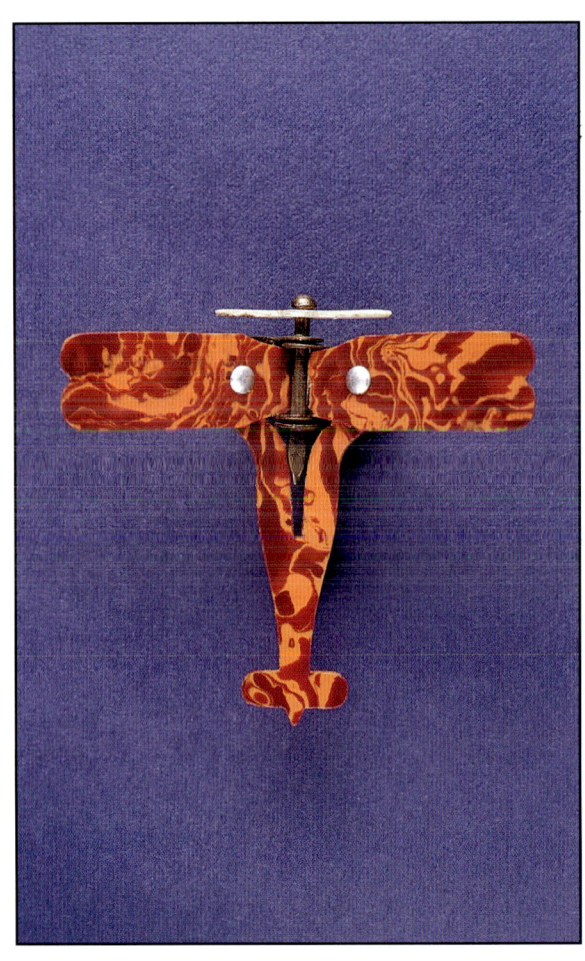
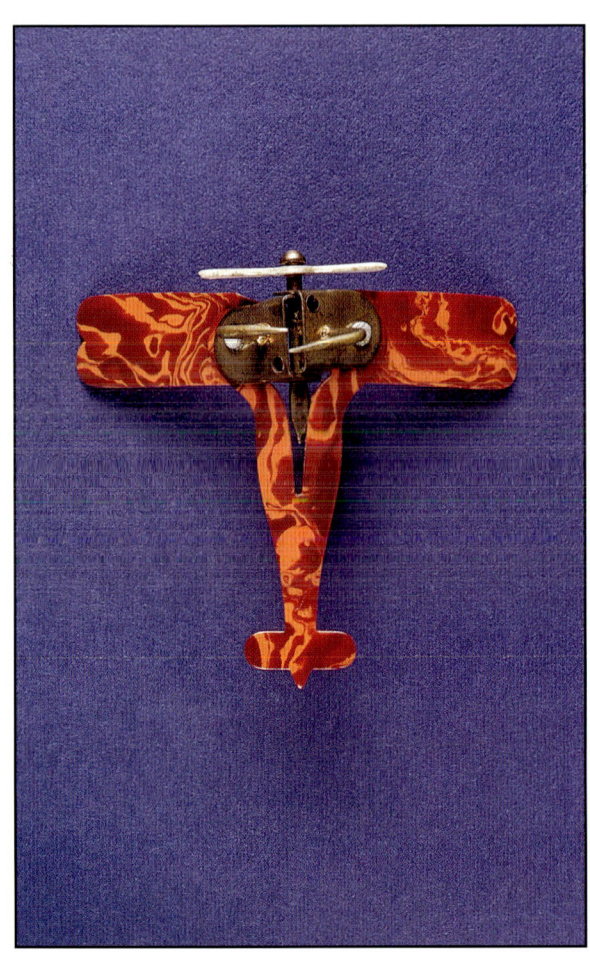

LEFT AND RIGHT
Airplane brooch made from a sheet of celluloid. $30-50.

LEFT AND RIGHT
Airplane brooch. $40-50

Elephant brooch, phenolic. 1930s. *Courtesy of Violet Nankervis.* $15-25.

Hatpins. $20-35 each.

Necklace and matching earrings made by John Maclellan using original materials. Celluloid.

Necklace and matching earrings made by John Maclellan using original materials. Celluloid.

OPPOSITE PAGE
Necklace and matching earrings made by John Maclellan using original materials. Celluloid.

Necklace and matching earrings made by John Maclellan using original materials. Phenolic.

Phenolic buckle and buttons. $45-50.

Phenolic brooch. $25-30.

Deco belt buckle. $60-65.

Celluloid on metal buckle, 1920s. $60-65.

Cleopatra belt buckle. $45-60.

Necklace and bracelet set. Green and amber phenolic. *Courtesy of Astoria.* $240-300.

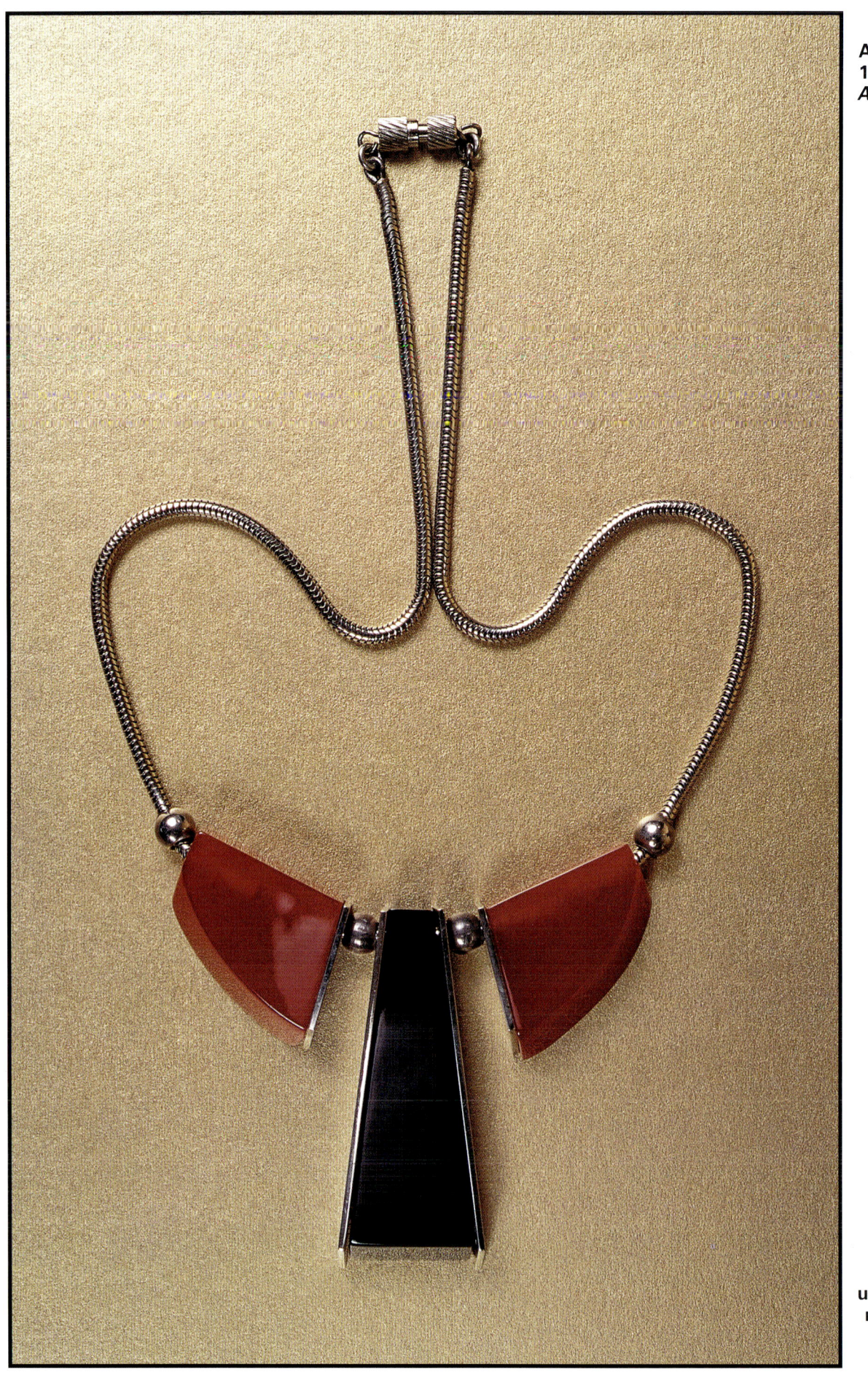

Art Deco necklace, 1930s. *Courtesy of Astoria.* $480-600.

OPPOSITE PAGE
Green and cream urea formaldehyde necklace on metal chain. $35-40.

Fabulous Art Deco necklace. *Courtesy of Astoria.* $480-560.

Necklace contains 'Opium' solid perfume. The design is in keeping with other objects in this book, but was actually produced in the 1970s. Material unknown. A good investment at $50-75.

Reticules.

These now rare items were carried by fashionable ladies in the 1920s. The beautiful bags contain a powder puff and mirror. The lipstick is often concealed in the tassle. Made of celluloid. Prices start at $240 and go up.

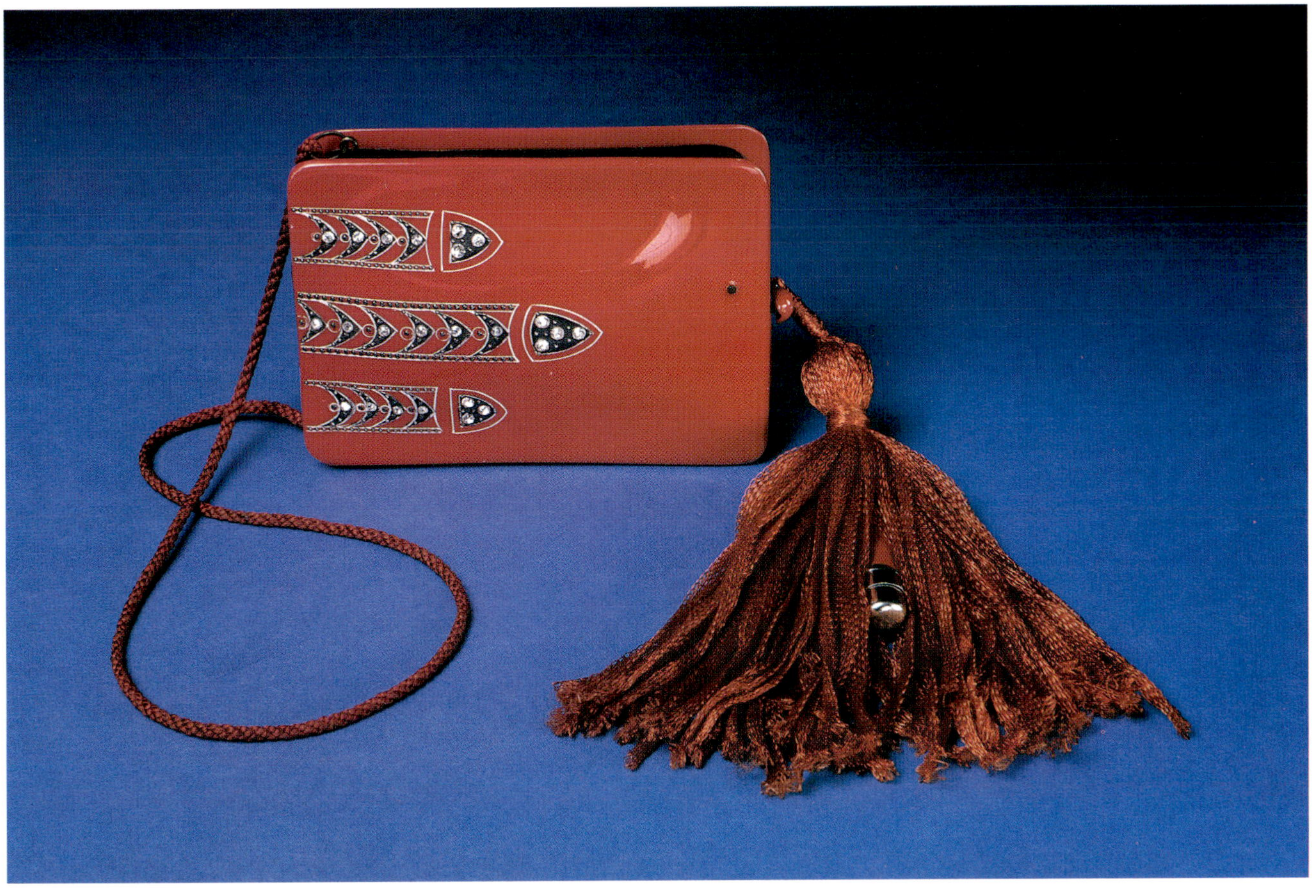

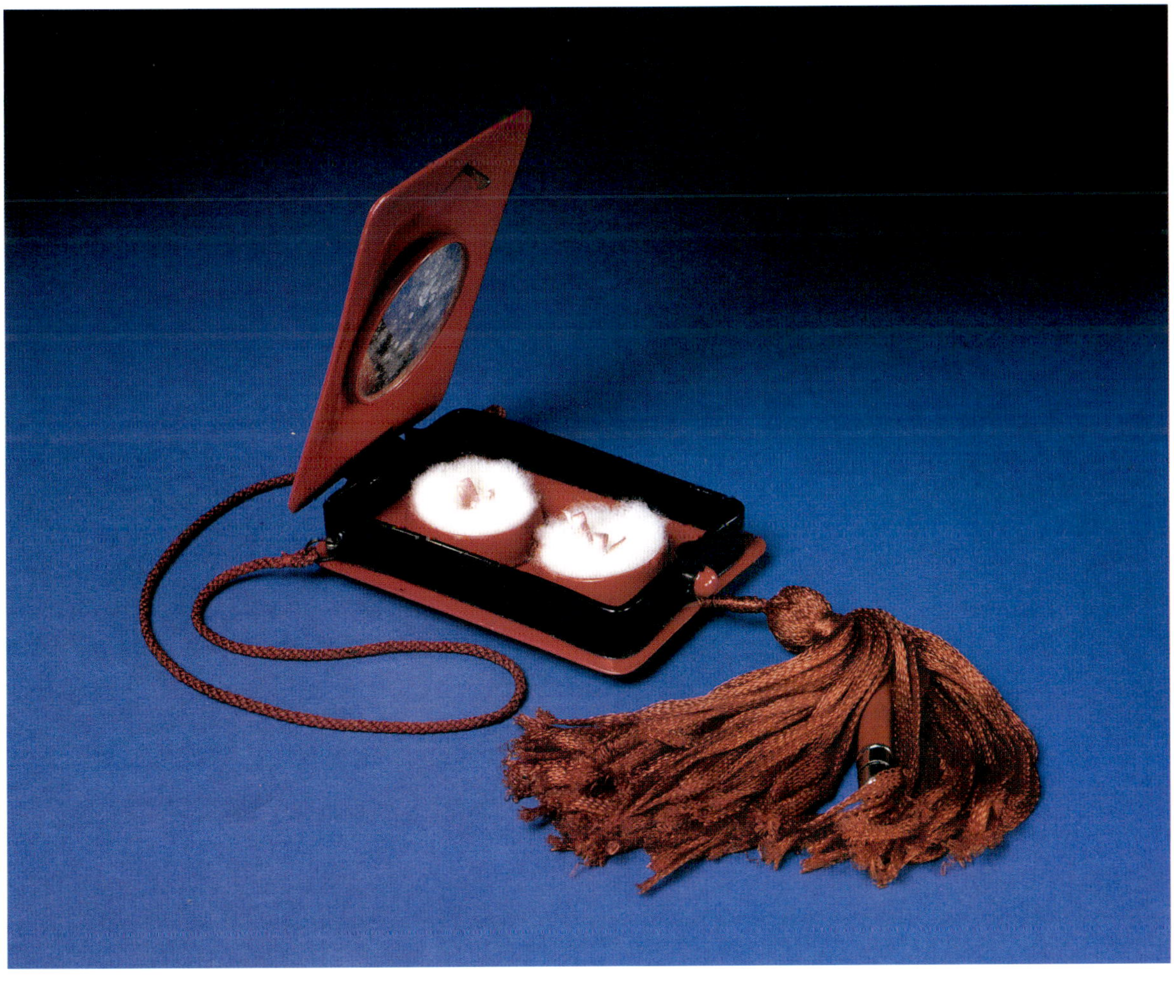

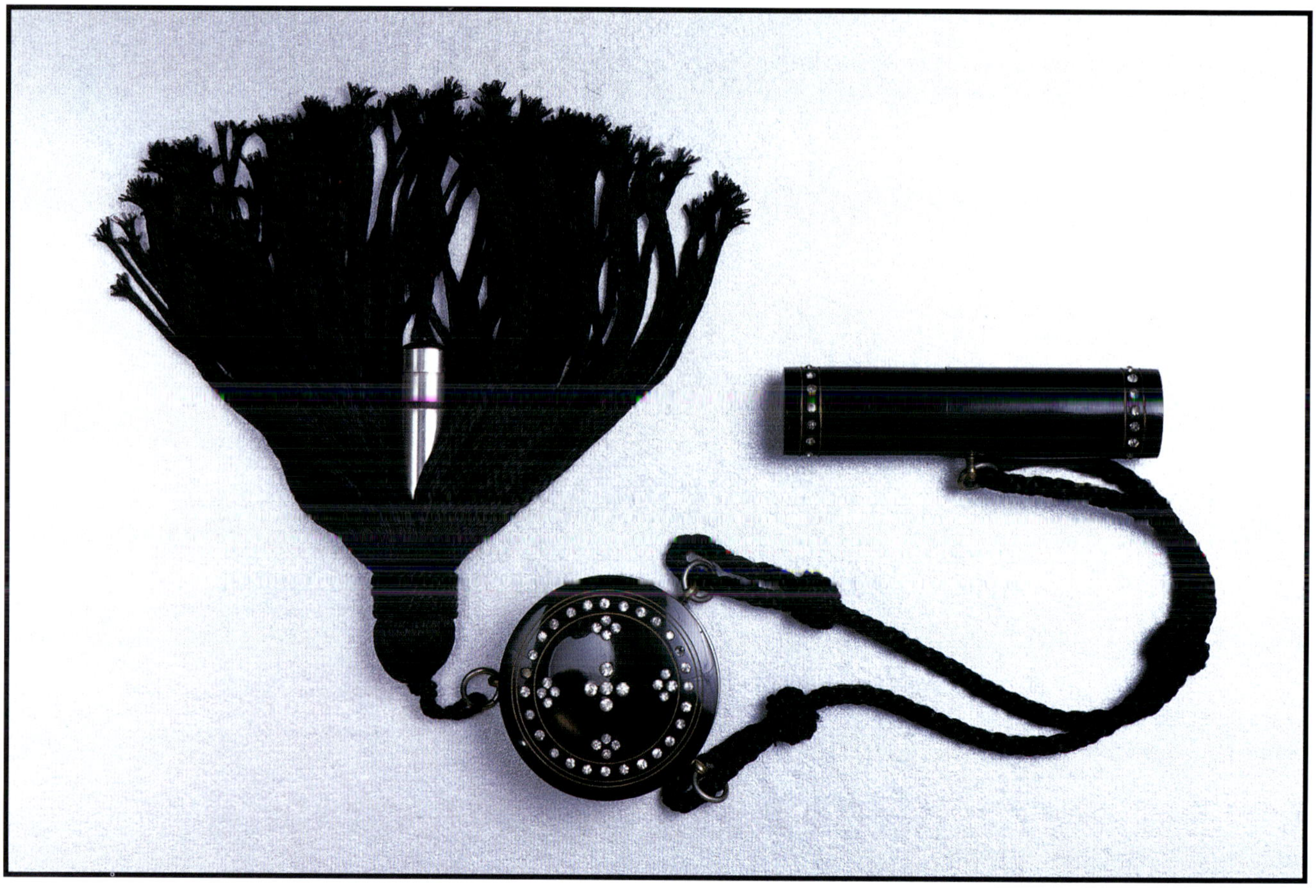

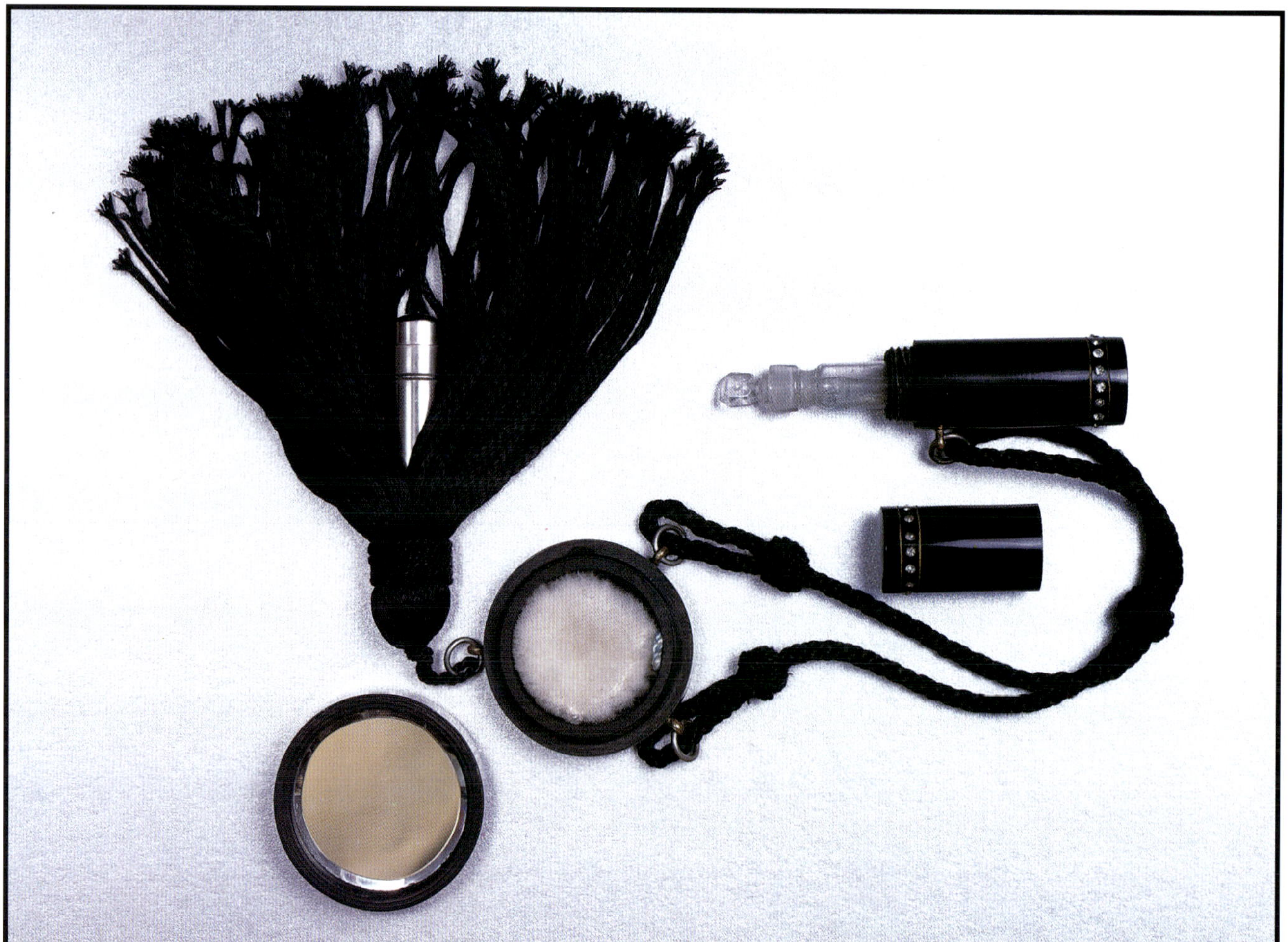

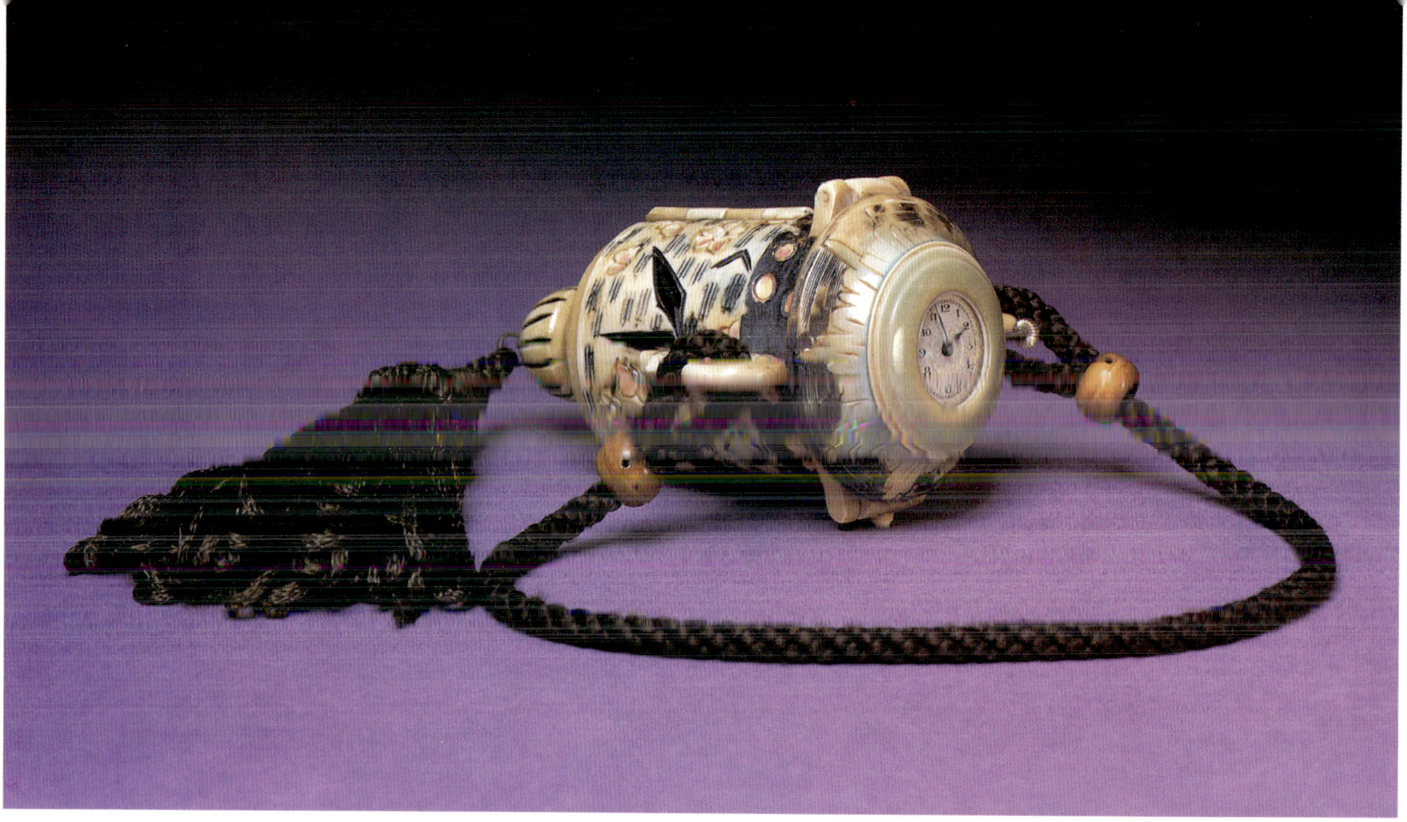

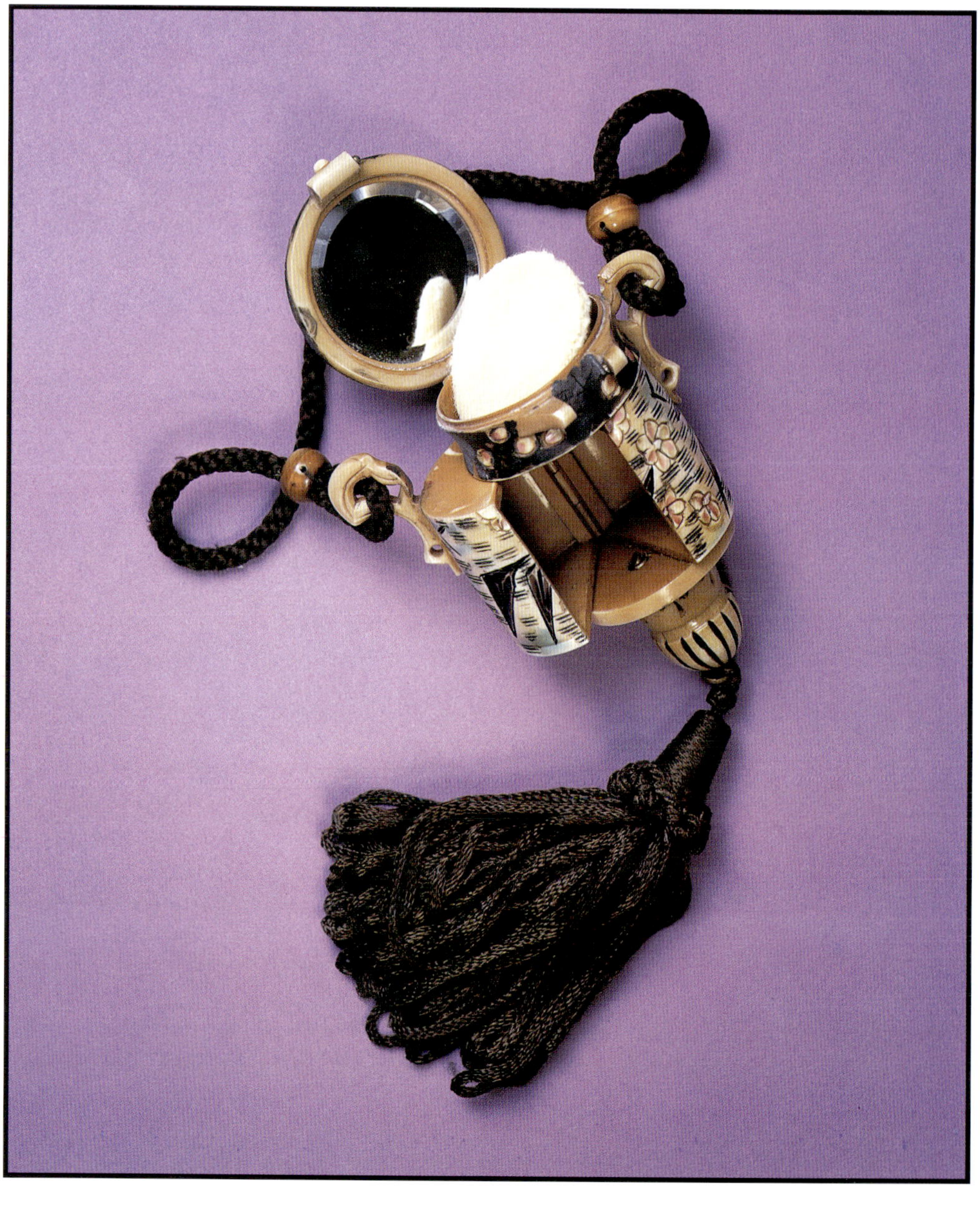

Celluloid handbag top. $150

Chapter 4.

"Grandma's Attic"

Toys and Miscellaneous

Enjoy this chapter, then roll up your shirtsleeves,
pull down the loft ladder, brush aside the cobwebs,
and search for your own long-lost treasures.

Doll's house radios and television made in Welwyn Garden City by Kleeware, whose customers included Woolworth stores. $8-12 each.

Baby's ball. Celluloid c1900. The parents among you may be interested to know that the ball rattles because it contains a handful of lead shot! $35-40.

Speckled phenolic child's or doll's tea set made in England, 1930s. (The doll's name is Mabel.) *Courtesy of Violet Nankervis.* $75-85.

Celluloid clown and donkey toy, joints are articulated. $75-85.

Battery operated nightlights, bases are made of cardboard. Only one has two faces. Celluloid. 1920s. $300-325 set.

Celluloid pencil sharpener. $35-40.

Pull on the cigar to reveal a tape measure. Celluloid. $125-160.

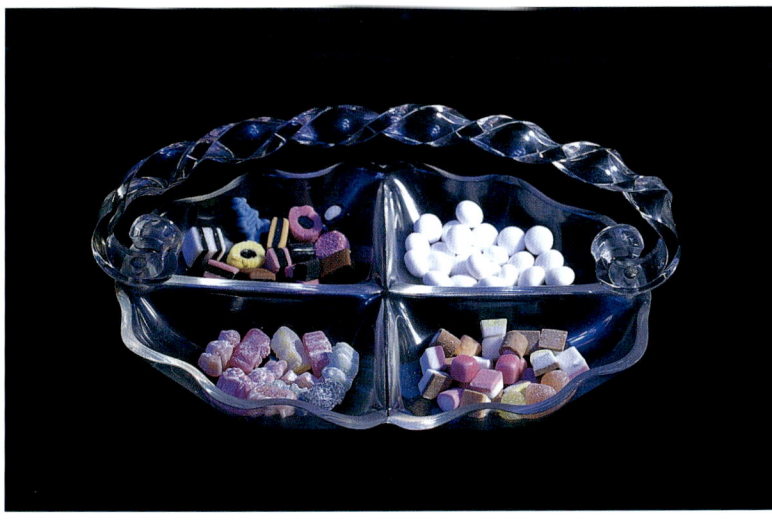

Acrylic dish with barley-twist handle. 1950s. *Photo by Brandon McGuire.* $30-35.

Spectacles. Possibly 1970s. I should think Dame
Edna would kill to own these. $185-225.

Ebena display stands. $75-85 set of three.

Nut bird, vulture. Phenolic and a brazil nut body, metal eyes. 3" high. $30-35.

Walking stick handle. Phenolic.
1930s. $75-85.

Freddie the Phenolic Frog
who started life as a
worthless prince. $65-80.

Celluloid fans. 1920s. $60-80 each.

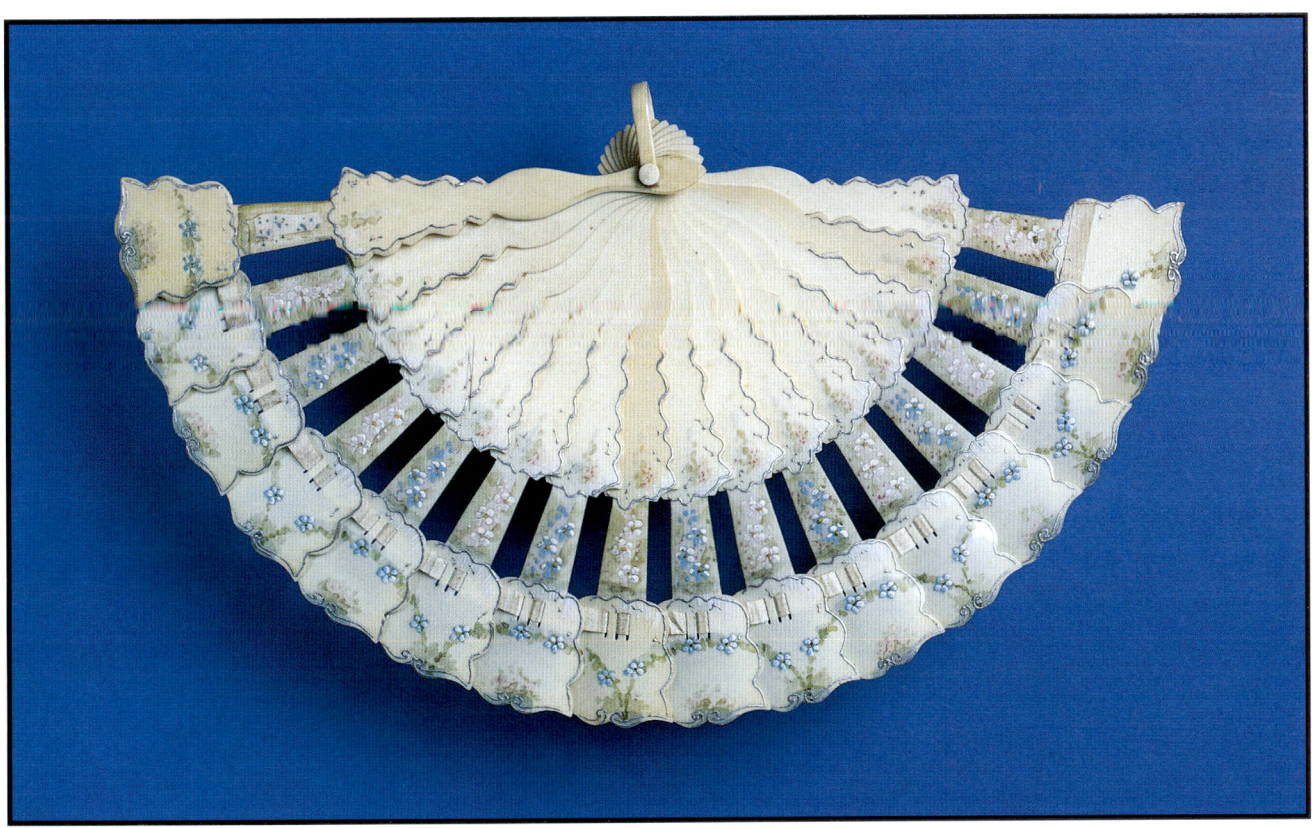

Appendix

"Some Candid Thoughts, One Collector To Another"

Where to buy

Antiques and Art Deco Fairs

These are specialist fairs where traders stall-out for the day to sell 20th century antiques. Many of the dealers are Bakelite fans and their stalls will usually contain several quality pieces. The London and international trade buy at these fairs, so come early for a better chance of securing something special.

These fairs take place several times a year. For times and dates, see the antiques trade journals in your region, such as the *Antique Trade Calendar*, or *BBC Homes and Antiques* magazine. Another reliable location to buy is Alfie's mall, Church Street, London. NW8. Nearest Tube is Edgeware Road.

Buying at Auction

This can be great fun, however, a few simple rules apply. Check the object carefully for damage (dirt can hide those hairline cracks). Register for the sale; you will probably be given a paddle or buyer's number and the auctioneer will take note of this if a lot is knocked down to you.

There are several ways to play the auction game. Some people just raise their hand and keep it there until the lot is theirs. This can prove expensive and I don't recommend it. Not every auctioneer is a paragon of virtue and you may find yourself bidding against the invisible man. Better to hang back and see how the bidding opens, then step in to join the bidding. Many people set themselves limits and of course limits are usually round numbers and you can often secure the lot with one extra bid.

A "Buyer's Premium," is a percentage added to the hammer price that can be as high as 17 ½%. Check the auction house's policy on this when you register to bid.

Plastics as an Investment

Any investment has an element of risk. Returns cannot be guaranteed. However, prices for Bakelite, cast phenolic, celluloid, etc. have risen by 100% (read it and weep, Wall Street) in the last five years.

Caring For Plastics

For most plastics a warm soapy wash will suffice. (Don't submerge the object for long periods.) Dry and buff with a soft cloth. For Bakelite and cast phenolic use metal polish (test a hidden area first) followed by a soapy wash. Dry and apply a little furniture polish. Then buff to required sheen. For celluloid wipe with a damp cloth and dry.

Some plastics degrade with age and with exposure to light, heat and damp. When displaying objects make sure there is free air flow and a reasonably consistent temperature. Display cases can cut out dusting but can increase the likelihood of degrading. I once saw a rare celluloid hair comb which

had started to degrade in a display case. It was covered in a light furry mould and if not removed it is quite likely it would have infected the surrounding objects. The worse case scenario of this would be that left unchecked you would end up with an empty display case!

How do I know if it's Bakelite?

1. Ask the person who is selling it. This works most of the time although this advice should only be used at specialist fairs and shops. The man at the antiques market who sells a broad selection of goods may really believe the R2D2 he's offering is indeed Bakelite.

2. Is the object in the style of the period, is it mottled or marbled and is the quality good? If it is unfilled phenolic are the colours strong and slightly translucent?

3. Still in doubt? Smell it. Breathe on the item, give it a quick rub and you will get a whiff of carbolic soap.

Where To See Great Plastics

The Bakelite Museum, Orchard Mill, Williton, Somerset. Tel. 01984 632133. Open from Easter to the end of September.
www.Bakelitemuseum.co.uk

This museum should be Number 1 on your 'must visit' list. Two floors of antique plastics set in the unlikely home of an old watermill. This collection has been put together by Bakelite guru, Patrick Cook. It's all here from an eggcup to a Bakelite caravan. Your visit will be further enhanced if you are lucky enough to have a chat with Patrick whose knowledge of the subject is vast.

For the overseas visitor, why not combine your visit with a trip to Stonehenge? Then stay over at a traditional English Bed & Breakfast.

The Design Museum, Butlers Wharf, London. SE1 2YD. Tel. 0171 403 6933. Open 11.30am - 6.00pm Mon-Fris. 10.30am - 6.00pm Sat - Suns. Nearest Tubes are Tower Hill, London Bridge, or Tower Gateway DLR.

Victoria and Albert Museum, South Kensington, London. SW7 2RL. Nearest Tube is South Kensington.

Science Museum, South Kensington, London. SW7 2DD. Nearest Tube is South Kensington.

Recommended Reading

Collins, Philip, *Smokerama, Classic Tobacco Accoutrements*, Chronicle Books, 1992.

Cook, Patrick and Catherine Slessor, *Bakelite*, Apple Press, 1992.

Dinoto, Andrea, *Art Plastic, Designed For Living*, Abberville Press, 1984.

Grasso, Tony, *Bakelite Jewellery, A Collector's Guide*, Apple Press, 1996

Katz, Sylvia, *Classic Plastics, From Bakelite to High Tech*, Thames and Hudson, 1984.

Quye, Anita and Colin Williamson, ed.*Plastics, Collecting and Conserving*, NMS Publishing Limited, 1999.

Glossary of Plastics Terms

Acrylic- plastic glass, better known by ICI's trade name Perspex and Lucite in America, available from the 1930s onward.

Bakelite- the first synthetic plastic, Phenol Formaldehyde, which was discovered by Belgian chemist Dr. Leo Baekeland while he was working in New York in 1907. Phenol Formaldehyde was patented in 1909 and known as the "heat and pressure patent." He then formed The General Bakelite Company in 1910. The new product, called Bakelite, was a big success and was used extensively in industry and the home. Bakelite has now become a generic term and is used to describe many types of plastics. Antique dealers use the word "Bakelite" to mean dark mottled and marbled plastics and the word "phenolic" to describe the more translucent, brighter colors. This is not always technically correct, but it works in the marketplace.

Bandalalsta- A line of dishes produced by Brookes and Adams using Urea and Thiourea Formaldehyde.

Beatl- Products made from beetle powder were displayed in Harrods in 1926 and sold by Selfridges soon after. Originally called Beetle, but this was changed to Beatl after complaints from customers who did not like the insect connotations. Note no such objections were heard from Woolworths, who also stocked this tableware.

Celluloid- Invented by John Wesley Hyatt in the late 1800s. A semi-synthetic plastic containing powdered camphor. A big success, it's uses included piano keys, dolls, dentures, shirt collars and of course, cine film. It's major drawback is that it is flammable.

Cellulose Acetate- A 1920s plastic which became a replacement for celluloid as it was non-flammable. Still in use today.

Ebonite- also known as hard rubber or Vulcanite. A very hard, black, heat-resistant thermoset plastic. Its uses include photographic equipment, telephones, and pipe stems.

Erinoid- Erinoid Limited was a company which produced Casein, a semi-synthetic plastic which contained milk protein. The name has become a generic term. Erinoid was successfully used to make many items, including knitting needles.

Linga Longa- A line of dishes produced by Brookes and Adams using Urea and Thiourea Formaldehyde.

Lnsden Ware- An English company that produced Bakelite mouldings.

Melamine- Often known as Formica or Warcite, it was used as tableware by US Airlines and kitchen worktops in England during the 1950s

Nylon- a synthetic first shown at the New York World's Fair in 1939 as bristles for tooth and hair brushes and, of course, nylon stockings.

Phenolic- A shortened version of "Phenol Formaldehyde," the first synthetic plastic which was discovered by Belgian chemist Dr. Leo Baekeland while he was working in New York in 1907. Until this time plastics had been natural (horn, shellac) or semi-synthetic (celluloid, casein).

Polythene- First produced by ICI around 1936, it was used during World War 2 in radar equipment and later for washing up bowls, polythene bags, etc. In 1951, a one-piece shuttlecock was exhibited at the Festival of Britain as a replacement for the traditional 23 pieces of leather, cork and goose feathers it usually consisted of.

PVC - Polyvinyl Chloride, used in bottles, wallpaper, and motorway cones, which I am sure we are all familiar with.

Thiourea Formaldehyde- an earlier and not so strong version of urea formaldehyde.

Urea Formaldehyde- Invented in the 1920s, this helped expand color variations including white, not before available. Brookes and Adams produced it's Bandalasta and Linga Longa ranges using Urea and Thiourea Formaldehyde.

Notes